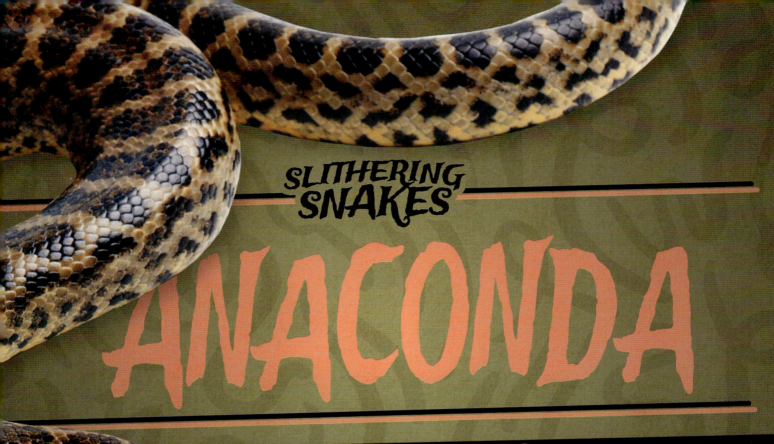

Slithering Snakes

ANACONDA

THE LARGEST SNAKE ON EARTH

BY NATALIE HUMPHREY

Enslow Publishing

DISCOVER!

Please visit our website, www.enslow.com. For a free color catalog of all our high-quality books, call toll free 1-800-398-2504 or fax 1-877-980-4454.

Library of Congress Cataloging-in-Publication Data

Names: Humphrey, Natalie, author.
Title: Anaconda : the largest snake on earth / Natalie Humphrey.
Description: New York : Enslow Publishing, [2021] | Series: Slithering snakes | Includes index.
Identifiers: LCCN 2019059391 | ISBN 9781978517653 (library binding) | ISBN 9781978517639 (paperback) | ISBN 9781978517646 (6 Pack) | ISBN 9781978517660 (ebook)
Subjects: LCSH: Anacondas—Juvenile literature.
Classification: LCC QL666.O63 H86 2021 | DDC 597.96/7—dc23
LC record available at https://lccn.loc.gov/2019059391

Published in 2021 by
Enslow Publishing
101 West 23rd Street, Suite #240
New York, NY 10011

Copyright © 2021 Enslow Publishing

Designer: Sarah Liddell
Editor: Natalie Humphrey

Photo credits: Cover, p. 1 (anaconda) cellistka/Shutterstock.com; background pattern used throughout Ksusha Dusmikeeva/Shutterstock.com; background texture used throughout Lukasz Szwaj/Shutterstock.com; p. 5 Morgan Paul/500px/Getty Images; pp. 7, 15 Vladimir Wrangel/Shutterstock.com; p. 9 Jeffdelonge/Wikimedia Commons; p. 11 Sylvain Cordier/Photodisc/Getty Images; p. 13 huang jenhung/Shutterstock.com; p. 17 Oradol/Shutterstock.com; p. 19 worldclassphoto/Shutterstock.com; p. 21 Volina/Shutterstock.com.

Portions of this work were originally authored by Johanna Burke and published as *Anaconda*. All new material this edition authored by Natalie Humphrey.

All rights reserved. No part of this book may be reproduced in any form without permission in writing from the publisher, except by a reviewer.

Printed in the United States of America

Some of the images in this book illustrate individuals who are models. The depictions do not imply actual situations or events.

CPSIA compliance information: Batch #BS20ENS: For further information contact Enslow Publishing, New York, New York, at 1-800-398-2504.

Find us on f ⊙

CONTENTS

Record-Breaking Snakes 4
Big Babies 8
Sneaky Serpent 10
Giant Anacondas 12
Yellow Anacondas 16
People and Anacondas 20
Words to Know 22
For More Information 23
Index. 24

Boldface words appear in Words to Know.

RECORD-BREAKING SNAKES

When people talk about giant snakes, the anaconda is always at the top of the list! There are four **species** of anaconda, but the most common are the green and the yellow anaconda. The green anaconda can grow to be the largest snake in the world!

GREEN ANACONDA

Anacondas live in South America. They like hot, wet places like the Amazon **rain forest**. They spend most of their time in the water and are very good swimmers. When anacondas are on land, they can even climb trees!

ANACONDAS ARE PART OF THE BOA FAMILY OF SNAKES.

BIG BABIES

Female anacondas are usually larger than males. In September through October, mother anacondas have around 20 to 50 babies, but they can have more! A baby green anaconda is about 2 feet (60 cm) long. Baby anacondas can hunt and swim right away.

MOTHER ANACONDAS DON'T TAKE CARE OF THEIR BABIES.

SNEAKY SERPENT

When an anaconda hunts, it first hides in the water to wait for its **prey**. When an animal walks by, the anaconda quickly bites it. Then the snake wraps its body around the prey. It **squeezes** the animal tight to kill it and swallows it whole!

Anacondas have curved teeth that hold prey tight!

11

GIANT ANACONDAS

The green anaconda, or giant anaconda, is the kind of anaconda that makes the snake famous. It can weigh over 550 pounds (250 kg) and grow to be 30 feet (9 m) long. This river monster is the heaviest snake in the world!

THE GREEN ANACONDA IS ALSO CALLED THE COMMON ANACONDA.

Anything that comes close to the water is a good meal for a green anaconda. It waits until an animal comes to drink. Anacondas eat deer, fish, birds, turtles, and even **caimans**! After killing its prey, the anaconda eats its meal in the water.

GREEN ANACONDAS SPEND MOST OF THEIR TIME IN THE WATER.

YELLOW ANACONDAS

The yellow anaconda is smaller than its green **relative**, but it's still a big snake! Yellow anacondas can be more than 12 feet (3.7 m) long and can weigh over 85 pounds (40 kg). They are found in Paraguay, Brazil, Argentina, and Bolivia.

Yellow anacondas hunt in the water like green anacondas, but they also hunt on land. Yellow anacondas eat birds, **rodents**, fish, turtles, and young caimans. Larger yellow anacondas can even take down wild pigs and small deer!

ANACONDAS EAT ONLY WHEN THEY'RE HUNGRY.

PEOPLE AND ANACONDAS

Anacondas would rather stay in their trees or in the water than go near people. In some places, people even keep yellow anacondas as pets! While there are still **legends** and stories about "man-eating" anacondas, none of them have been found to be true.

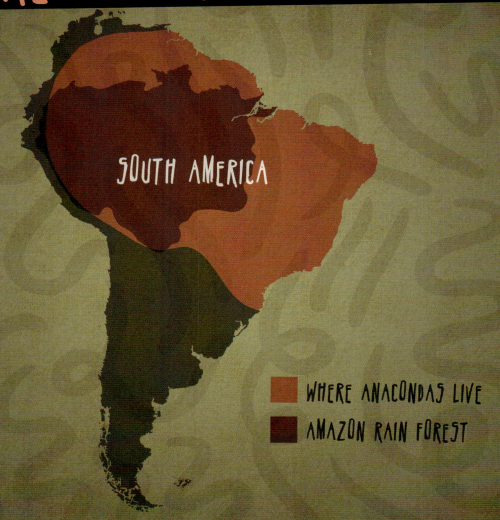

WORDS TO KNOW

caiman A type of small alligator.
legend A story from the past that is believed by many people but cannot be proven to be true.
prey An animal that is hunted or killed by another animal for food.
rain forest A forest that receives a lot of rain and has very tall trees.
relative Something that belongs to the same group as something else because of shared characteristics.
rodent A small animal (such as a mouse, rat, or squirrel) that has sharp front teeth.
species A group of animals that are similar and can produce young animals.
squeeze To press together tightly.

FOR MORE INFORMATION

Books

Golkar, Golriz. *Anacondas*. Minneapolis, MN: Cody Koala, 2018.

Lawrence, Ellen. *Green Anaconda*. Minneapolis, MN: Bearport Publishing, 2017.

Websites

National Geographic Kids
kids.nationalgeographic.com/animals/reptiles/anaconda/
Check out more fun facts and maps of where anacondas live!

San Diego Zoo Kids
kids.sandiegozoo.org/animals/anaconda
With interactive graphics, dive into more fun anaconda facts!

Publisher's note to educators and parents: Our editors have carefully reviewed these websites to ensure that they are suitable for students. Many websites change frequently, however, and we cannot guarantee that a site's future contents will continue to meet our high standards of quality and educational value. Be advised that students should be closely supervised whenever they access the internet.

INDEX

baby anacondas, 8

green anacondas, 4, 8, 12, 14, 16, 18

hunting, 8, 10, 14, 18

"man-eating" anacondas, 20

pet anacondas, 20

prey, 10, 14, 18

rain forest, 6

size, 8, 12, 16

species of anaconda, 4

water/swimming, 6, 8, 10, 14, 18, 20

weight, 12, 16

where they are found, 6, 16

yellow anacondas, 4, 16, 18, 20